After twelve years of persistent conflict, the United States finds itself in a familiar situation — facing a declining defense budget and a strategic landscape that continues to evolve. As our current large-scale military campaign draws down, the United States still faces a complex and growing array of security challenges across the globe as "wars over ideology have given way to wars over religious, ethnic, and tribal identity; nuclear dangers have proliferated; inequality and economic instability have intensified; damage to our environment, food insecurity, and dangers to public health are increasingly shared; and the same tools that empower individuals to build enable them to destroy."[2] Unlike past draw downs, where the threats we faced were going away, there remain a number of challenges that we still have to confront -- challenges that call for a change in America's defense priorities. Despite these challenges, the United States Army is committed to remaining capable across the spectrum of operations. While the future force will become smaller and leaner, its great

---

[1] The Posture of the United States Army, Committee on Armed Services, United States House of Representatives, April 23, 2013.
[2] National Security Strategy, May 2010.

strength will lie in its increased agility, flexibility, and ability to deploy quickly, while remaining technologically advanced. We will continue to conduct a complex set of missions ranging from counterterrorism, to countering weapons of mass destruction, to maintaining a safe, secure and effective nuclear deterrent. We will remain fully prepared to protect our interests and defend our homeland.[3]

The Army depends on its Science and Technology (S&T) program to help prepare for the future, mitigate the possibility of technical surprise and ensure that we remain dominant in any environment. The Army's S&T mission is to foster discovery, innovation, demonstration and transition of knowledge and materiel solutions that enable future force capabilities and/or enhance current force systems. The Army counts on the S&T Enterprise to be seers of the future – to make informed investments now, ensuring our success for the future.

The Army is ending combat operations in Afghanistan and refocusing on the Asia-Pacific region with greater emphasis on responses to sophisticated, technologically proficient threats. We are at a pivotal juncture – one that requires us to relook the past twelve years of conflict and capitalize on all the lessons that we have learned, while we implement a strategic shift to prepare for a more capable enemy. As the Department of Defense prepares for the strategic shift, the Army will adapt — remaining an ever present land force — unparalleled throughout the World.

We are grateful to the members of this Committee for your sustained support of our Soldiers, your support of our laboratories and centers and your continued commitment to ensure that funding is available to provide our current and future Soldiers with the technology that enables them to defend America's interests and those of our allies around the world.

Strategic Landscape

As we built the FY15 President's Budget Request, the Army faced a number of significant challenges. While the Army has many priorities, the first and foremost priority is and always will be to support our Soldiers in the fight. We are pulling our troops and equipment out of Afghanistan by the end of this December, we are drawing down our force structure, we are resetting our equipment after 12 plus years of war and we are trying to modernize. Given the budget downturn within the Department of Defense, the Army has been forced to face some

---

[3] "The Posture of the United States Army," The Honorable John M. McHugh, Secretary of the Army and General Raymond T. Odierno, Chief of Staff, United States Army before the Senate Committee on Appropriations, Subcommittee on Defense, May 22, 2013.

difficult choices. The Army is in the midst of a significant force structure reduction – taking the Army to pre-World War II manning levels. The Chief of Staff of the Army has undertaken difficult decisions balancing force structure, operational readiness, and modernization to maintain a capable force able to prevent, shape and win in any engagement. As a result, over the next five years, we face a situation where modernization will be slowed, new programs will not be initiated as originally envisioned and the Army's S&T Enterprise will be challenged to better prepare for the programs and capabilities of the future. We will focus on maturing technology, reducing program risk, developing prototypes that can be used to better define requirements and conducting experimentation with Soldiers to refine new operational concepts. The S&T community will be challenged to bring forward not only new capabilities, but capabilities that are affordable for the Army of the future.

> *"Going forward, we will be an Army in transition. An Army that will apply the lessons learned in recent combat as we transition to evolving threats and strategies. An Army that will remain the best manned, best equipped, best trained, and best led force as we transition to a leaner, more agile force that remains adaptive, innovative, versatile and ready as part of Joint Force 2020."*[4]
>
> — *General Raymond T. Odierno, 38th Chief of Staff, Army*

Goals and Commitments

The emerging operational environment presents a diverse range of threats that vary from near-peer to minor actors, resulting in new challenges and opportunities. In this environment, it is likely that U.S. forces will be called upon to operate under a broad variety of conditions. This environment requires a force that can operate across the range of military operations with a myriad of partners, simultaneously helping friends and allies while being capable of undertaking independent action to defeat enemies, deter aggression, and shape the environment. At the same time, innovation and technology are reshaping this environment, multiplying and intensifying the effects that even minor actors are able to achieve.

The Army's S&T investment is postured to address these emerging threats and capitalize on opportunities. The S&T investment continues to not only focus on developing more capable and affordable systems, but also on understanding the complexity of the future environment. We have focused on assessing technology

---

[4] "Marching Orders," General Raymond T. Odierno, 38th Chief of Staff, U.S. Army, January 2012.

and system vulnerabilities (from both a technical and operational perspective) to better effect future resilient designs and to prepare countermeasures that restore our capabilities when necessary.

There are persistent (and challenging) areas where the Army invests its S&T resources to ensure that we remain the most lethal and effective Army in the world. As the Army defines its role in future conflicts, we are confident that these challenges will remain relevant to the Army and its ability to win the fight. The S&T community is committed to help enable the Army achieve its vision of an expeditionary, tailorable, scalable, self-sufficient, and leaner force, by addressing these challenges:

- Enabling greater *force protection* for Soldiers, air and ground platforms, and bases (e.g., lighter and stronger body armor, helmets, pelvic protection, enhanced vehicle survivability, integrated base protection)
- *Easing overburdened* Soldiers in small units (both cognitive and physical burden, e.g., lighter weight multi-functional materials)
- Enabling *timely mission command and tactical intelligence* to provide situation awareness and communications in ALL environments (mountainous, forested, desert, urban, jamming, etc.)
- *Reducing logistic burden* of storing, transporting, distributing and retrograding materials
- Creating operational overmatch (*enhancing lethality and accuracy*)
- Achieving *operational maneuverability* in all environments and at high operational tempo (e.g., greater mobility, greater range, ability to operate in high/hot environments)
- Enabling *early detection and treatment for* Traumatic Brain Injury *(TBI) and* Post Traumatic Stress Disorder *(PTSD)*
- *Improving operational energy* (e.g., power management, micro-grids, increased fuel efficiency engines, higher efficiency generators, etc.)
- Improving *individual and team training* (e.g., live-virtual-constructive training)
- *Reducing lifecycle costs* of future Army capabilities

In addition to these enduring challenges, the S&T community conducts research and technology development that impacts our ability to maintain an agile and ever ready force. This includes efforts such as establishing environmentally compatible installations and materiel without compromising readiness or training, creating leader selection methodologies, and new test tools that can save resources and reduce test time, and establishing methods and measures to improve Soldier and unit readiness and resilience.

The Army S&T strategy acknowledges that we must respond to the new fiscal environment and changing technology playing field. Many critical technology breakthroughs are being driven principally by commercial and international concerns. We can no longer do business as if we dominate the technology landscape. We must find new ways of operating and partnering. We realize that we should invest where the Army must retain critical capabilities but reap the benefits of commercially driven technology development where we can. No matter the source, we will ensure the Army is aware of the best and most capable technologies to enable a global, networked and full-spectrum joint force in the future. As the U.S. rebalances its focus by region and mission, it must continue to make important investments in emerging and proven capabilities. **In a world where all have nearly equal access to open technology, innovation is the most important discriminator in assuring technology superiority.**

The Chief of Staff of the Army has made his vision clear.

> *"The All-Volunteer Army will remain the most highly trained and professional land force in the world. It is uniquely organized with the capability and capacity to provide expeditionary, decisive landpower to the Joint Force and ready to perform across the range of military operations to Prevent, Shape, and Win in support of Combatant Commanders to defend the Nation and its interests at home and abroad, both today and against emerging threats.[5]"*
>
> *— General Raymond T. Odierno, 38th Chief of Staff, Army*

The Army is relying on its S&T community to carry out this vision for the Army of the future.

Implementing New Processes

Turning science into capability takes a continuum of effort including fundamental research, the development and demonstration of technology, the validation of that technology and its ultimate conversion into capability. From an S&T materiel perspective, this includes the laboratory confirmation of theory, the demonstration of technical performance, and the experimentation with new technologies to identify potential future capabilities and to help refine/improve system designs. But the S&T Enterprise is also charged with helping to

---

[5] Gen Raymond Odierno, 38th Chief of Staff Army, "CSA Strategic Priorities, Waypoint 2", 2014

conceptualize the future -- to use our understanding of the laws of physics and an ability to envision a future environment to broaden the perspective of the requirements developers as well as the technology providers.

As part of this continuum, the Army has adopted a 30 year planning perspective to help facilitate more informed program planning and budget decisions. A major part of the S&T strategy is to align S&T investments to support the acquisition Programs of Record (PoRs) throughout all phases of their lifecycle and across the full DOTMLPF (Doctrine, Organization, Training, Materiel, Leadership, Personnel, and Facilities). By expanding the perspective, areas where there are unaffordable alignments of activities (such as multiple major Engineering Change Proposals in the same portfolio within the same 2-3 year timeframe) or unreasonable alignments (such as planned technology upgrades to a system that has already transitioned into sustainment) are made obvious. With that information in mind, the Army has established "tradespace" to generate options that inform strategic decisions that allow the Army to stay within its fiscal top line while maximizing its capabilities for the Warfighter.

This new and ongoing process, known as the Long Range Investment Requirements Analysis (LIRA), has put additional rigor into the development of the Army's budget submission and creates an environment where the communities who invest in all phases of the materiel lifecycle work together to maximize the Army's capabilities over time. From an S&T perspective, it clearly starts to inform the materiel community as to WHEN technology is needed for insertion as part of a planned upgrade. It also cues us as to when to start investing for replacement platforms. In addition, this long-range planning can introduce opportunities for convergence of capabilities such as the development of a single radar that can perform multiple functions for multiple platforms or the convergence of cyber and Electronic Warfare (EW) capabilities into one system. Aside from the obvious benefit achieved by laying out the Army's programs and seeing where we may have generated unrealizable fiscal challenges, it has reinvigorated the relationships and strengthened the ties between the S&T community and their Program Executive Office (PEO) partners. We are working together to identify technical opportunities and the potential insertion of new capabilities across this 30 year timeframe.

The LIRA process was used to inform the development of the FY15 President's Budget. As the Army faced a dramatic decline in its modernization accounts (a 40% decrement over the next two years), we used the results of the LIRA to ensure that we had a fiscally sound strategy.

The S&T Portfolio

The nature of Science and Technology is such that continuity and stability have great importance. Starting and stopping programs prevents momentum in research and lengthens the timelines for discovery and innovation. While the Army S&T portfolio gains valuable insight from the threat community, this only represents one input to the portfolio and likely describes the most probable future. To have a balanced outlook across all the possible futures requires that the portfolio also address the "possible" and "unthinkable." The Army's S&T portfolio is postured to address these possible futures across the eight technology portfolios identified Figure 1.

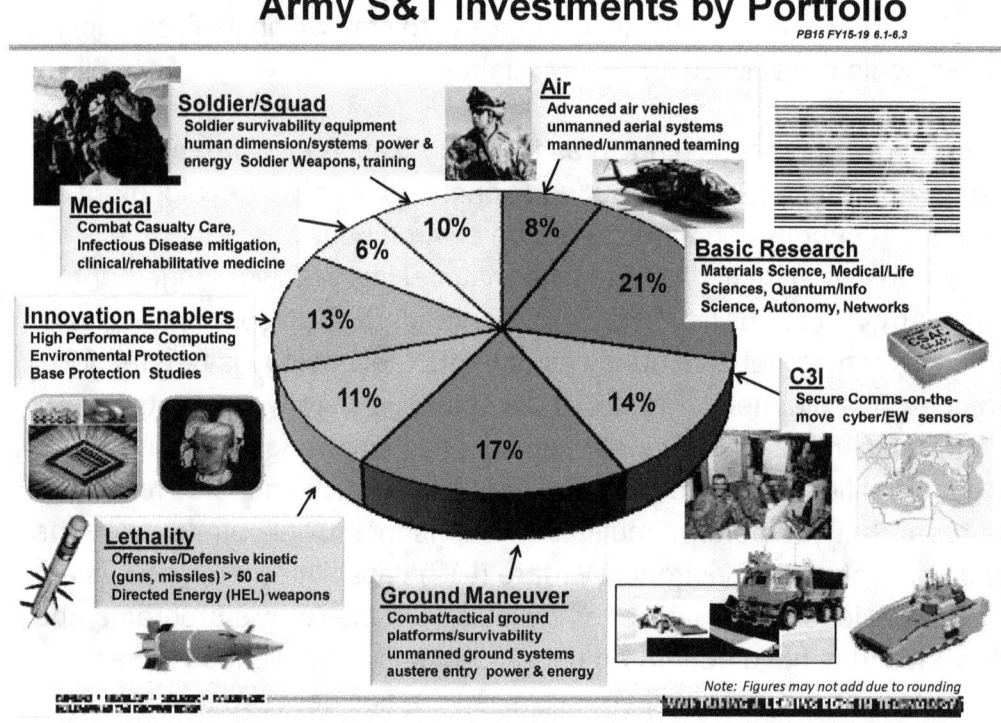

Figure 1. Army S&T Investments by Portfolio

The efforts of the S&T Enterprise are managed by portfolio to ensure maximum synergy of efforts and reduction of unnecessary duplication. The S&T program is organized into eight investment portfolios that address challenges across six Army-wide capability areas (Soldier/Squad; Air; Ground Maneuver; Command, Control, Communications, and Intelligence (C3I); Lethality; and Medical) and two S&T enabling areas (Basic Research and Innovation Enablers).

The 2014 Quadrennial Defense Review (QDR) protects and prioritizes key investments in technology to maintain or increase capability while forces grow leaner. This is an opportunity to look at innovative applications of technology. As a result, in the FY15 President's Budget Request, the Army is maintaining, and shifting where necessary, its emphasis on technology areas that enable the Army to be leaner, expeditionary, and more lethal.

We are now in an era of declining acquisition budgets and are mindful of the challenges this brings to our S&T programs. We will have fewer opportunities for transition to Programs of Record in the next few years. This "pause" in acquisition does however afford us the opportunity to further develop and mature technologies, ensuring that when acquisition budgets do recover, S&T will be properly positioned to support the Army's next generation of capabilities. This year finds the Army beginning to rebalance its S&T funding between Basic Research, applied research and advanced technology development. We appreciate the flexibility that was provided to the DoD S&T executives to better align our funding to our Service/Agency needs after years of proscriptive direction.

In FY15, our Advanced Technology Development investments increase to 42% of our $2.2B budget. This is a deliberate increase from previous years as the Army looks to its S&T community to conduct more technology demonstration/ prototyping initiatives that will inform future Programs of Record (PoRs). Specifically you will see the Army shifting or increasing emphasis on research areas that support the next generation of combat vehicles (including power and energy efficiency, mobility and survivability systems), Anti-Access/Area Denial (A2/AD) technologies such as assured Position Navigation and Timing (PNT) and austere entry capabilities, Soldier selection tools and training technologies, as well as long range fires. Two of these efforts, the Future Infantry Fighting Vehicle (FIFV) and the assured Position Navigation and Timing (PNT) efforts are being done in collaboration with the respective PEOs to ensure that the capability developed and demonstrated not only helps to refine the requirements for the future PoRs but establishes an effective link for transition. We are also increasing our investments in vulnerability assessments of both technology and systems as well as expanding our Red Teaming efforts to identify potential vulnerabilities in emerging technologies, systems and systems-of-systems, including performance degradation in contested environments, interoperability, adaptability, and training/ease of use. This year begins the re-alignment necessary to implement our strategy of investing in areas critical to the Army – areas where we have critical skills sets, and leveraging others (sister services, other government agencies, academia, industry, allies) for everything else.

We anticipate a future where rapidly advancing technologies such as autonomous systems, high yield energetics, immersive training environments, alternative power and energy solutions, and the use of smart phones and social media will become critical to military effectiveness. The Army will continue to develop countermeasures to future threat capabilities and pursue technological opportunities. Enemies and adversaries however, will counter U.S. technological advantages through cover, concealment, camouflage, denial, deception, emulation, adaptation, or evasion. Finally, understanding how humans apply technology to gain capabilities and train will continue to be at least as important as the technologies themselves.

We are mindful however that the Army will continue to be called on for missions around the globe. The Army is currently deployed in ~160 countries conducting missions that range from humanitarian support to stability operations to major theater warfare. As we have seen in the last month, the world is an unpredictable place, and our Soldiers must have the capabilities to deal with an ever changing set of threats.

S&T Portfolio Highlights

I'd like to highlight a few of our new initiatives and remind you of some of our on-going activities that will help frame the options for the Army of the future.

Soldier/Squad Portfolio

One of the important initiatives currently underway that we anticipate will make major inroads into our efforts to lighten the Soldier's load is the development of a Soldier Systems Engineering Architecture. This architecture, developed in concert with our acquisition and requirements community, is an analytical decision-based model through which changes in Soldier system inputs (loads, technology/equipment, physiological & cognitive state, stress levels, training, etc.) may be assessed to predict changes in performance outputs of the Soldier system in operationally relevant environments. By using a systems engineering approach, the model will result in a full system level analysis capable of predicting impacts of both materiel and non-materiel solutions on fully equipped Soldiers performing operational missions/tasks

In keeping with the CSA's vision, our S&T efforts also support the Army's training modernization strategy by developing technologies for future training environments that sufficiently replicate the operational environment. We are also developing new training effectiveness measures and methods, ensuring that these new training technologies can rapidly and effectively transfer emerging warfighting experience and knowledge into robust capabilities. In addition, the need to reduce force structure has increased the importance of our research in the area of personnel selection and classification. This research will provide the Army with methods to acquire and retain candidates best suited for the Army – increasing our flexibility to adapt to changes in force size, structure and mission demands. Other important research includes developing scientifically valid measures and metrics to assess command climate and reduce conduct related incidences, including sexual harassment and assault in units to ensure the Army can maintain a climate of dignity, respect and inclusion.

Air Portfolio

As the lead service for rotorcraft, owning and operating over 80% of the Department of Defense's vertical lift aircraft, the preponderance of rotorcraft technology research and development takes place within the Army. Our key initiative, the Joint Multi-Role Technology Demonstrator (JMR TD) program, is focused on addressing the A2/AD need for longer range and more efficient combat profiles. As we shift to the Pacific Rim focus, future Areas of Operation (AO) may be sixteen times larger than those of our current AOs. The Army needs a faster, more efficient rotorcraft, capable of operating in high/hot environments (6000 feet and 95 degrees) with significantly decreased operating costs and maintenance required. The new rotorcraft will also require improved survivability against current and future threats. The goal of the JMR TD effort is to reduce risk for the Future Vertical Lift planned PoR, the Department of Defense's next potential "clean sheet" design rotorcraft. The overall JMR TD effort will use integrated government/industry platform design teams and exercise agile prototyping approaches. At the same time, the Army is collaborating with DARPA on their x-plane effort. While the DARPA program is addressing far riskier technologies that are not constrained by requirements, we will look to leverage technology advancements developed under the DARPA effort where possible.

Another initiative that we are beginning in FY15 is addressing one of the biggest causes of aircraft loss - accidents that occur while operating in a Degraded Visual Environments (DVE). DVE is much more than operating while in brown out – this effort looks at mitigating all sources of visual impairment, either those caused by the aircraft itself (brownout, whiteout) or other "natural" sources (rain, fog, smoke, etc.). We are currently conducting a synchronized, collaborative effort with PEO Aviation to define control system, cueing, and pilotage sensor combinations which enable maximum operational mitigation of DVE. This S&T effort will result in a prioritized list of compatible, affordable DVE mitigation technologies, and operational specification development that will help inform future Army decisions. This program is tightly coupled with the PEO Aviation strategy and potential technology off-ramps will be transitioned to the acquisition community along the way, when feasible.

Ground Maneuver Portfolio

The Ground Maneuver Portfolio is focused on maturing and demonstrating technologies to enable future combat vehicles, including the Future Infantry Fighting Vehicle (FIFV). In FY15, you will see the beginning of a focused initiative done in collaboration with PEO Ground Combat Systems, to develop critical sub-system prototypes to inform the development and requirements for the Army's Future Infantry Fighting Vehicle (FIFV). These sub-system demonstrators focus on mobility (e.g., engine, transmission, suspension); survivability (e.g., ballistic protection, under-body blast mitigation, advanced materials); Active Protection Systems (APS); a medium caliber gun and turret; and an open vehicle power and data architecture that will provide industry with a standard interface for integrating communications and sensor components into ground vehicles.

Armor remains an Army-unique challenge and we have persistent investments for combat and tactical vehicle armor, focusing not only on protection but also affordability and weight reduction. We continue to invest in advanced materials and armor technologies to inform the next generation of combat and tactical vehicles.

In FY15, this portfolio continues its shift to focus to address A2/AD challenges. We've increased efforts on technologies to enable stand-off evaluation of austere ports of entry and infrastructure to better enable our ability to enter areas of conflict. We are also maintaining technology investments in detection and neutralization of mines and improvised explosive devices to ensure freedom of maneuver.

C3I Portfolio

The C3I portfolio provides enabling capability across many of the challenges, but specifically seeks to provide responsive capabilities for the future in congested Electro-Magnetic environments. These capabilities are supported by sustained efforts in sensors, communications, electronic warfare and information adaptable in dynamic, congested and austere (disconnected, intermittent and limited) environments to support battlefield operations and non-kinetic warfare. Renewed efforts in the C3I portfolio include reinvigorating efforts in sensor protection. We continue to invest in EW vulnerability analysis to perform characterization and analysis of radio frequency devices to develop detection and characterization techniques, tactics, and technologies to mitigate the effects of contested environments (such as jamming) on Army C4ISR systems.

Given the potential challenges that we face while operating in a more contested environment, we are placing additional emphasis on assured Position, Navigation and Timing, developing technologies that allow navigation in Global Positioning System (GPS) denied/degraded environments for mounted and dismounted Soldiers and unmanned vehicles such as exploiting signals of opportunity. We will study improvements for high sensitivity GPS receivers that could allow acquisition and tracking in challenging locations such as under triple canopy jungles, in urban areas, and inside buildings. We are developing Anti-Jam capabilities as well as supporting mission command with interference source detection, signal strength measurement, and with locating interference sources, thereby enabling the Army to conduct its mission in challenging electromagnetic environments.

The C3I Portfolio also includes our efforts in cyber, both defensive and offensive. Defensive efforts in cyber security will investigate and develop software, algorithms and devices to protect wireless tactical networks against computer network attacks. We are developing sophisticated software assurance algorithms to differentiate between stealthy life cycle attacks and software coding errors, as well as investigating and assessing secure coding methodologies that can detect and self-correct against malicious code insertion. We will research and design sophisticated, optimized cyber maneuver capabilities that incorporate the use of reasoning, intuition, and perception while determining the optimal scenario on when to maneuver, as well as the ability to map and manage the network to determine probable attack paths and the likelihood of exploitation.

On the offensive side of cyber operations, we will develop integrated electronic attack (EA) and computer network operations hardware and software to execute force protection, EA, electronic surveillance and signals intelligence missions in a dynamic, distributed and coordinated fashion.

We will demonstrate protocol exploitation software and techniques that allow users to remotely coordinate, plan, control and manage tactical EW and cyber assets; develop techniques to exploit protocols of threat devices not conventionally viewed as cyber to expand total situational awareness by providing access to and control of adversary electronic devices in an area of operations.

Lethality Portfolio

In FY15, you will see continued emphasis on the development of A2/AD capabilities through Long Range Fires and Counter Unmanned Aircraft technologies. S&T is focusing on advanced seeker technologies to enable acquisition of low signature threats at extended ranges, along with dual pulse solid rocket motor propulsion to provide longer range rockets and extend the protected areas of air defense systems. To support these capabilities, we are conducting research in new energetic materials focused on both propulsive and explosive applications. These materials have significantly higher energetic yield than current materials and will increase the both effectiveness of our systems and reduce their size.

We also continue to develop Solid State High Energy Lasers to provide low cost defeat of rockets, artillery, mortars and unmanned aircraft. We have had multiple successes in High Energy Lasers, as we demonstrated successful tracking and defeat of mortars and unmanned aircraft in flight this year (FY14) from our mobile demonstrator.

Additionally, we are supporting the Ground Maneuver Portfolio in the demonstration of a medium caliber weapon system to enable Future Infantry Fighting Vehicle requirements for range and lethality including an airburst munition.

Medical

The Medical portfolio addresses the wellness and fitness of our Soldiers from accession through training, deployment, treatment of injuries and return to duty or to civilian life. Ongoing efforts address multiple threats to our Soldiers' health and readiness. Medical research focuses on areas of physiological and psychological health that directly support the Chief of Staff of the Army Ready and Resilience Campaign and the Army Surgeon General's Performance Triad (Activity, Nutrition and Sleep). Research in these portfolios includes important areas such as Traumatic Brain Injury (TBI) and Post Traumatic Stress Disorder (PTSD). In FY15, you will see continued focus on research to mitigate infectious diseases prevalent in the Far East as well as combat casualty care solutions at the point of injury that will extend Soldier's lives during the extended distances associated with conducting operations in the Pacific.

TBI research efforts include furthering our understanding of cell death signals and neuroprotection mechanisms, as well as identifying critical thresholds for secondary injury comprising TBI. The Army is also evaluating other non-traditional therapies for TBI, and identifying "combination" therapeutics that substantially mitigate or reduce TBI-induced brain damage. Current Army funded research efforts in the area of PTSD are primarily focused upon development of pharmacologic solutions for the prevention and treatment of PTSD. A large-scale clinical trial is currently underway evaluating the effectiveness of Sertraline, one of two Selected Serotonin Reuptake Inhibitors (SSRIs) approved for the treatment of civilian PTSD, but not combat-related PTSD. This study will evaluate Sertraline's effectiveness in the treatment of combat-related PTSD both alone and in combination with psychotherapy.

Innovation Enablers

As the largest land-owner/user within the DoD, it is incumbent upon the Army to be good stewards in their protection of the environment. As such, the Army develops and validates lifecycle models for sustainable facilities, creates dynamic resource planning/management tools for contingency basing, develops decision tools for infrastructure protection and resiliency and assesses the impact of sustainable materials/systems on the environment.

In addition, we conduct blast noise assessment and develop mitigation technologies to ensure that we remain "good neighbors" within Army communities and work to protect endangered species while we ensure that the Army mission can continue.

The High Performance Computing (HPC) Modernization Program supports the requirements of the DoD's scientists and engineers by providing them with access to supercomputing resource centers, the Defense Research and Engineering Network (DREN) (a research network which matures and demonstrates state of the art computer network technologies), and support for software applications, including the experts that help to improve and optimize the performance of critical common DoD applications programs to run efficiently on advanced HPC systems maturing and demonstrating leading-edge computational technology.

The Army's Technology Maturation Initiatives effort, established in FY12 enables a strategic partnership between the S&T and acquisition communities. This effort has become especially important as the Army heads into a funding downturn. We plan to use these funds to prepare the Army to capitalize on S&T investments as we come out of the funding "bathtub" near the end of the decade. We are using these Budget Activity 4 resources to target areas where acquisition programs intended to provide necessary capabilities have been delayed, such as assured Position, Navigation and Timing (PNT), the Future Infantry Fighting Vehicle and Active Protection Systems. We are investing resources that will either provide capability or inform/refine requirements for the Army's future systems (all of which will be done via collaborative programs executed with our acquisition/PEO partners).

This portfolio includes our ManTech efforts as well. Last month, President Obama announced the launch of the Digital Manufacturing and Design Innovation Institute (DMDI). Headquartered in Chicago, Illinois, and managed by the U.S Army's Aviation and Missile Research Development and Engineering Center, the DMDI Institute spearheads a consortium of 73 companies, universities, nonprofits, and research labs. The president announced a government investment of $70 million and matching private investments totaling $250 million for the institute. DMDI is part of the president's National Network of Manufacturing Innovation (NNMI) and will focus on the development of novel model-based design methodologies, virtual manufacturing tools, and sensor and robotics based manufacturing networks that will accelerate the innovation in digital manufacturing and increase U.S. competitiveness.

Basic Research

Underpinning all of our efforts and impacting all of the enduring Army challenges is a strong basic research program. Army Basic Research includes all scientific study and experimentation directed toward increasing fundamental knowledge and understanding in those fields of the physical, engineering, environmental, and life sciences related to long-term national security needs. The vision for Army Basic Research is to advance the frontiers of fundamental science and technology and drive long-term, game-changing capabilities for the Army through a multi-disciplinary portfolio teaming our in-house researchers with the global academic community to ensure overwhelming land-warfighting capabilities against any future adversary.

While we have made some significant adjustments within the Basic Research investments within the Army, we will continue to emphasize several areas that we feel have a high payoff potential for the Warfighter. These areas include: Materials in Extreme Environments; Quantum Information and Sensing; Intelligent Autonomous Systems; and Human Sciences/Cybernetics.

For centuries, the fabrication of solid materials has hinged largely on manipulating a narrow range of temperatures and pressures. Our Materials in Extreme Environments initiative invests in new revolutionary and targeted scientific opportunities to discover and exploit the fundamental interaction of matter under extreme static pressures and magnetic fields, controlled electromagnetic wave interactions (microwave, electrical) and acoustic waves (ultrasound) to dramatically enhance fabrication and create engineered materials with tailored microstructures and revolutionary functionalities.

We are in the midst of a second quantum revolution – moving from merely computing quantum properties of systems to exploiting them. Areas of particular focus for the Army include quantum enhanced sensing and imaging, quantum communications, quantum algorithms, and quantum simulations. For example, an Army-specific quantum-enabled capability is an exact polynomial-time quantum-chemistry algorithm that directly impacts the design of propellants, explosives, medicines, and materials.

To enable the Warfighter, animal-like intelligence is desired for simple autonomous platforms, such as robotic followers, and for aerial and ground sensor platforms. We are investing in research that will enable highly intelligent systems that allow platforms to set waypoints autonomously, increasing mission effectiveness; followers that recognize the actions of their unit, that can perceive when the unit is deviating from a previously prescribed plan and know enough to query why; and that recognize when the unit is resting and be capable of doing so without explicit instructions from the Soldier.

Regardless of specific definition, human sciences are critical and can safely be predicted to become pervasive across all Army research activities. Cognitive predictions of social person-to-person communication based on observed gestures, eye movement, and body language are becoming possible. In addition, brain-to-brain interaction is emerging as a potential paradigm based on external sensors and brain stimulation. The Army will continue to study these and other possible techniques, to understand shared knowledge, social coordination, discourse comprehension, and detection and mitigation of conflict. Cognitive models combined with sensors also have the potential for dramatic breakthroughs in human-autonomy interaction, including aspects such as active learning algorithms, real-time crowd-sourcing with humans and machines in the cloud, and maximizing AI prediction accuracy. Devices and sensors that are wearable or implantable (including biomarkers and drug therapy) have the potential to enhance performance dramatically and to augment sensory information through new human-sensor-machine interface designs.

The role of Basic Research to provide the knowledge, technology, and advanced concepts to enable the best equipped, trained and protected Army to successfully execute the national security strategy, cannot be understated. The key to success in Basic Research is picking the right research challenges, the right people to do the work, and providing the right level of resources to maximize the likelihood of success.

Impact of Sequestration

I am often asked what impact sequestration had on the Army's S&T portfolio, so I would like to address some of the impacts we have seen. The FY13 application of sequestration targets (hitting every Program Element in the S&T portfolio by a set percentage) forced the Army into a scenario where we decremented programs that we would have protected, if given the opportunity. This lack of flexibility made for some very bad business and technical ramifications. Within the S&T community, we were able to balance our sequestration targets at the Program Element, vice Project level – giving us the ability to avoid civilian Reduction in Force (RIF) actions where possible. That said, sequestration did result in unfunded efforts and delays in applied research and technology development areas across the S&T portfolio. More generally, the sequestration cuts added unnecessary risk to acquisition programs and delayed the transition of critical capabilities to the Warfighter.

However, by far the most serious consequence of sequestration (and the related pay freezes, shutdowns, conference restrictions, etc.) has been the impact on our personnel. Without a world-class cadre of scientists and engineers, the Army S&T enterprise would be unable to support the needs of the Army. The Army Labs and Research, Development and Engineering Centers have reported multiple personnel leaving for other job opportunities or early retirement. For example, the Night Vision and Electronic Sensors Directorate lost eight personnel in the two months prior to the well-publicized DoD-wide furloughs, compared to an average annual loss of around 19 personnel. These losses include personnel across experience levels with specialized expertise critical to the Army. While the average attrition rate over the past two years is running at about 8% (similar to a typical attrition rate found in prior years), the concerning impact is that 60% of the personnel leaving the Army are NOT eligible for retirement. This is a big change. During our exit interviews, reasons cited included conference restrictions (impeding the ability to progress professionally) coupled with increasing job insecurity due to budget decrements and planned manpower reductions. Complicating this loss of technical expertise is the restriction on hiring replacements for the lost government civilians. We are on a replacement cycle that varies between 1 hire per every 3 losses at one lab, to 1 hire for every 20 losses at another. This pattern of loss is unsustainable if we hope to maintain a premier technical workforce. Finally, as we address the 2013 National Defense Authorization Act (NDAA), Section 955 language which mandates a reduction in the civilian workforce commensurate with a reduction in the military, we must confront the impacts of any civilian reductions, which are implemented through a personnel process that tends to primarily impact those employees who have less tenure in the government. For the S&T community that typically impacts those areas of new technical emphasis within the DoD – key areas such as cyber research and systems biology.

While the Bipartisan Budget Act has provided some relief and stability for FY14 and FY15, the uncertainty again looming on the horizon makes it even more difficult to recruit and retain the scientists and engineers the Army depends on. As you know, the key to any success within the Army lies with our people.

The S&T Enterprise Infrastructure and Workforce

Our laboratory infrastructure is aging, with an average approximate age of 50 years. Despite this, the S&T Enterprise manages to maximize the scarce sustainment, restoration, and modernization funding and the authorities for minor military construction using NDAA, Sec. 219 funding to minimize the impact on the R&D functions with the Enterprise. However, we are only making improvements to our infrastructure at the margins, and where possible we have used MILCON, through your generous support and unspecified minor construction to modernize facilities and infrastructure. However, we do acknowledge that much of the Army is in a similar position. This is not a long-term solution. While the authorities that you have given us have been helpful, they alone are not enough, and we are still faced with the difficulty of competing within the Army for ever-scarcer military construction dollars at the levels needed to properly maintain world-class research facilities. This will be one of our major challenges in the years to come and I look forward to working with OSD and Congress to find a solution to this issue.

The S&T community affords us the flexibility and agility to respond to the many challenges that the Army will face. Without the world-class cadre of over 12,000 federal civilian scientists and engineers and the infrastructure that supports their work, the Army S&T Enterprise would be unable to support the needs of the Army. To maintain technological superiority now and in the future, the Army must maintain an agile workforce. Despite this current environment of unease within the government civilian workforce, exacerbated by conference restrictions, budget uncertainty, furloughs, and near zero pay increases, we continue to have an exceptional workforce. But, as I mentioned earlier, attracting and retaining the best science and engineering talent into the Army Laboratories and Centers is becoming more and more challenging. Our laboratory personnel demonstrations give us the flexibility to enhance recruiting and afford the opportunity to reshape our workforce, and I appreciate Congress' continued support for these authorities to include the flexibilities given to the Laboratories and Centers in the 2014 NDAA, Section 1107 language. With two exceptions (the Army Research

Institute (ARI) for the Behavioral and Social Sciences and the Space and Missile Defense Command Technical Center (SMDCTC)), all of our laboratories and centers are operating under this program (ARI and SMDCTC were never designated as Science and Technology Reinvention Laboratories). The flexibilities given to the laboratories and centers allow the laboratory directors the maximum management flexibility to shape their workforce and remain competitive with the private sector.

The Army S&T Enterprise cannot survive without developing the next generation of scientists and engineers. We continue to have an amazing group of young scientists and engineers that serve as role models for the next generation. For example, last year Dr. Ronald Polcawich, a researcher at the U.S. Army Research Laboratory (ARL), was named by President Obama to receive a 2012 Presidential Early Career Award for Scientists and Engineers as one of the nation's outstanding young scientists for his work in Piezoelectric-Micro Electro-Mechanical Systems (PiezoMEMS) Technology. Dr. Polcawich, is leading a team of researchers at the ARL in studying PiezoMEMS with a focus on developing solutions for RF systems and actuators for millimeter-scale robotics. These actuators combined/integrated with low power sensors are being developed to enable mm-scale mechanical insect-inspired robotic platforms.

The need for STEM literacy, the ability to understand and apply concepts from science, technology, engineering and mathematics in order to solve complex problems, goes well beyond the traditional STEM occupations of scientist, engineer or mathematician. The Army also has a growing need for highly qualified, STEM-literate technicians and skilled workers in advanced manufacturing, logistics, management and other technology-driven fields. Success and sustainment for the Army S&T Enterprise depends on a STEM-literate population to support innovation and the Army must contribute to building future generations of STEM-literate and agile talent.

Through the Army Educational Outreach Program (AEOP), the Army makes a unique and valuable contribution to meet the national STEM challenge - a challenge which includes the growing demand for STEM competencies; the global competitiveness for STEM talent; an unbalanced representation of our nation's demographics in STEM fields; and the critical need for an agile and resilient STEM workforce. AEOP offers a cohesive, collaborative portfolio of STEM programs that provides students, as well as teachers, access to our world-class Army technical professionals and research centers. Exposure to STEM fields and STEM professionals is critical to growing the next generation of STEM-literate young men and women who will form the Army's workforce of tomorrow.

In the 2012-2013 academic year, AEOP directly engaged more than 66,000 students and nearly 1,500 teachers in authentic research experiences. Almost 2,351 Army Scientists and Engineers (S&E's) provided mentorship, either from our in-house research laboratories or through our university partnerships. Additionally in FY13, we initiated a comprehensive evaluation strategy (the first of its kind) that uses the government and a consortium of STEM organizations known for their nationwide education and outreach efforts to annually assess our program. Aligned with Federal guidance, AEOP requires the evaluation of all elements of the program based on specific, cohesive, metrics and evidence-based approaches to achieve key objectives of Army outreach; increased program efficiency and coherence; the ability to share and leverage best practices; as well as focus on Army priorities. The AEOP Priorities are:

- STEM Literate Citizenry: Broaden, deepen and diversify the pool of STEM talent in support of the Army and our defense industry base.

- STEM Savvy Educators: Support and empower educators with unique Army research and technology resources.

- Sustainable Infrastructure: Develop and implement a cohesive, coordinated and sustainable STEM education outreach infrastructure across the Army.

For FY15, we are concentrating on implementing evidence-based program improvements, strengthening additional joint service sponsored efforts, and identifying ways to expand the reach and influence of successful existing programs by leveraging partnerships and resources with other agencies, industry and academia.

New Approaches to Enhance Innovation

It is widely acknowledged that innovation depends on bringing multiple scientific disciplines together to engage in collaborative projects -- often yielding unpredictable, yet highly productive results. Formal and informal interactions among scientists lead to knowledge-building and research breakthroughs. These types of collaborations are happening on a day-to-day basis across our labs and engineering centers to produce the superior technology that our Army needs today, tomorrow and beyond. With shrinking budgets and huge leaps in the pace of technological change, our Army science and technology organizations must do more with less and faster than ever before to develop technology that will ensure mission success for the Army's first battle after next. To this end, we must more succinctly leverage scientific discovery from our academic and industry base by increasing the scientific engagement and flow of ideas that leads to ground breaking innovation.

In 1945, Vannevar Bush's concepts documented in "Science - the Endless Frontier" stressed the necessity of a robust/synergistic university, industry and government laboratory research system. Over the years, the rigid and insular nature of the defense laboratories have caused an erosion of that university/industry/government lab synergy that is critical to the discovery, innovation and transition of science and technology important to national security.

In an effort to reenergize that synergy, the US Army Research Laboratory (ARL) is working to extend their alliances through an Open Campus Concept that brings together under one roof the triad of industry, academia, and government. Leveraging the cutting-edge innovation of academia, the system development and transition expertise of industry and their own Army-focused fundamental research; ARL can harness the power of the triad to produce revolutionary science and technology more efficiently and effectively. The Open Campus Concept creates an ecosystem for academia, defense labs, and industry to share people, facilities and resources to develop and deliver transformative science oriented on solving complex Army problems. It will provide the means for our world-class scientific talent to work together in state-of-the-art facilities to provide innovation that allows rapid transition of technology to our Soldiers. ARL's Open Campus Concept could lead to a new business model that would transform the defense laboratory enterprise into an agile, efficient and effective laboratory system that supports the continuous flow of people and ideas to ensure transformative scientific discovery, innovation and transition critical to national security.

Finally, we are increasingly mindful of the globalization of S&T capabilities and expertise. Our International S&T strategy provides a framework to leverage cutting edge foreign science and technology enabled capabilities through Global Science and Technology Watch, engagement with allies and leadership initiatives. Global S&T Watch is a systematic process for identifying, assessing, and documenting relevant foreign research and technology developments. The Research, Development and Engineering Command's (RDECOM) International Technology Centers (ITCs), Engineer Research and Development Center (ERDC) international research office and the Medical Research Materiel Command's OCONUS laboratories identify and document relevant foreign S&T developments. We have initiated a new process to strategically identify and selectively engage our allies when their technologies and materiel developments can contribute to Army needs and facilitate coalition interoperability. The resultant engagements will augment the existing bilateral leadership forums we currently maintain with the United Kingdom Canada, Germany and Israel which provide both visibility of and management decisions on allied developments that merit follow-up for possible collaboration.

Summary

As the Army S&T program continues to identify and harvest technologies suitable for transition to our force, we aim to remain ever vigilant of potential and emerging threats. We are implementing a strategic approach to modernization that includes an awareness of existing and potential gaps; an understanding of emerging threats; knowledge of state-of-the-art commercial, academic, and government research; as well as a clear understanding of competing needs for limited resources. Army S&T will sharpen its research efforts to focus upon those core capabilities it needs to sustain while identifying promising or disruptive technologies able to change the existing paradigms of understanding. Ultimately, the focus remains upon Soldiers; Army S&T consistently seeks new avenues to increase the Soldier's capability and ensure their technological superiority today, tomorrow, and decades from now. The Army S&T mission is not complete until the right technologies provide superior, yet affordable, overmatch capability for our Soldiers. I will leave you with a last thought from the Secretary of the Army, the Honorable John McHugh.

> *"Our Strategic Vision is based on a decisive technological superiority to any potential adversary."*[6]
>
> *— Honorable John W. McHugh, 21st Secretary of the Army*

This is an interesting, yet challenging, time to be in the Army. Despite this, we remain an Army that is looking towards the future while taking care of the Soldiers today. I hope that we can continue to count on your support as we move forward, and I would like to again thank the members of the Committee for all you do for our Soldiers. I would be happy to take any questions you have.

---

[6] Terms of Reference, FY12 Army Science Board Summer Study, Secretary of the Army, John M. McHugh, October 28, 2011.

www.ingramcontent.com/pod-product-compliance
Lightning Source LLC
Chambersburg PA
CBHW081824170526
45167CB00008B/3538